PLANETARY EXPLORATIO

JUPITER

Don Davis and Carolyn Collins Petersen

Series editor: Dr David Hughes

Facts On File
New York • Oxford

Facts On File, Inc.
460 Park Avenue South
New York NY 10016
USA

Library of Congress Cataloging-in-Publication Data

Petersen, Carolyn Collins.
    Jupiter / text, Carolyn Collins Petersen ; illustrations, Donald
Davis.
        p.   cm. -- (Planetary exploration)
    Includes index.
    ISBN 0-8160-2048-5 (alk. paper)
    1. Jupiter (Planet)--Juvenile literature.   2. Project Voyager-
Juvenile literature.   I. Title.   II. Series.
QB661.P46   1989   89-31686
                                                          CIP
                                                                AC

Facts On File books are available at special discounts when
purchased in bulk quantities for businesses, associations,
institutions or sales promotion. Please contact the Special
Sales Department of our New York office at 212/683-2244
(dial 800/322-8755 except in NY, AK or HI).

Designed and produced by BLA Publishing Limited,
East Grinstead, Sussex, England.

A member of the **Ling Kee Group**
LONDON · HONG KONG · TAIPEI · SINGAPORE · NEW YORK

Phototypeset in Britain by BLA Publishing/Composing Operations
Color origination in Hong Kong
Printed and bound in Portugal

10 9 8 7 6 5 4 3 2 1

Note to the reader
In this book some words are printed in **bold** type.
These words are listed in the glossary on page 42.
The glossary gives a brief explanation of words which
may be new to you.

Contents

Foreword

Long before man had the ability to explore space, he imagined what it might be like. Now that our journey beyond the Earth's atmosphere has begun, an artist can draw upon scientific knowledge gathered by probes and satellites in space, and observations made here on Earth, to portray a world where man has never been. Using what scientists know about the violent beginning of our solar system, the artist can take us back into the past to witness the formation of the planets, or into the future to imagine how they might one day be colonized. Through these paintings we can dive into the clouds of Jupiter, hover above the furnace of a sunspot, or even look back on our own solar system as we travel farther away into the galaxy.

In the six volumes of the **Planetary Exploration** series, we have combined the most advanced knowledge about the planets of our solar system with the extraordinary work of a noted space artist. Each book, written by an expert in the field, takes the reader beyond current facts and theories to the frontier of the unknown: the surface of Mars, the rings of Saturn, the tiny glacial world of Pluto, the many moons of Uranus, and beyond. Artist Don Davis has matched these exciting scientific discoveries with vivid illustrations that allow us to "travel" to these planets and unlock their mysteries.

We hope that, in **Planetary Exploration**, you will enjoy sharing this adventure.

David W. Hughes

A giant planet

Jupiter is a giant planet wrapped in clouds that reflect sunlight, making it shine brightly in the sky. The colored patterns of the clouds produce graceful swirls across the **disc** and in the planet's southern hemisphere, a giant whirling spot peers out like an angry red eye. Dozens of smaller white and yellow spots give the planet an alien appearance. On the night side of the planet, lightning flashes light up the clouds with an eerie glow.

A swarm of satellites orbit Jupiter, like the planets orbiting around our Sun, and a thin ring of dust particles encircles it like a halo.

▼ Jupiter's orbit is 5.2 times bigger than Earth's, taking 11.86 years to complete. Jupiter is by far the biggest planet in the solar system, with a mass 317 times that of Earth and a diameter 11 times greater.

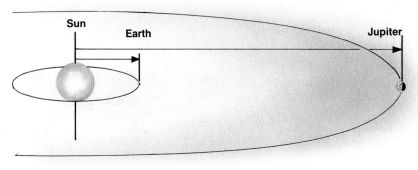

Sun   Earth   Jupiter

▶ We look down here onto Jupiter's cloud-covered north pole. The bright bands on the planet are known as "zones" and the darker regions between the zones are called "bands." Like the other giant planets, Jupiter spins quickly and all the bands and zones run parallel to the equator. The rapid motion also makes the planet **oblate**; the polar diameter is 83,400 miles (134,200 km), while the equatorial diameter is larger — 88,732 miles (142,796 km).

The north polar region is dusky and featureless. The two widest, orange-colored bands on each side of the equator are known as the north and south equatorial belts. The Great Red Spot can be seen at the lower edge of the planetary disc, at latitude 22° south. The spot is 16,300 miles (26,200 km) long by 8,600 miles (13,800 km) wide.

The clouds spin at slightly different rates, the polar clouds taking 5 minutes 11 seconds longer to go around the planet than the equatorial ones.

Aurorae and lightning can be seen illuminating parts of Jupiter's night side.

Facts about Jupiter

Diameter	88,732 miles (142,796 km)	**Inclination of equator**	3.08°
Mass	317 times the Earth's	**Surface gravity**	2.3 times the Earth's
Distance from the Sun	483 million miles (778 million km)	**Temperature**	−240°F −150°C
Period of rotation	9.841 hours	**Atmosphere**	90 % hydrogen, 10 % helium
Orbital period	11.86 years	**Number of moons**	more than 16

Above the cloud tops

◄ The Great Red Spot can be seen at top left. To the south is a mass of swirling, spiraling clouds. The bluish regions are "canyons" — great chasms that have opened up between the clouds.

In the orange-red cloud belts, the deepest regions are the darkest in color. By looking at these dark swirls, we are, in effect, descending into the depths of Jupiter's churning interior. The white clouds are the highest although, at times, veils of orange haze drift above them.

Remember that the higher clouds are colder than the lower clouds, and all the regions shown have a temperature that is far below the freezing point of water.

Jupiter is the largest planet in our solar system. It is the fifth farthest from the Sun, orbiting at a distance of almost 500 million miles (800 million km). By comparison, Earth orbits the Sun at only 94 million miles (150 million km). At times during their orbits, Earth and Jupiter are about 400 million miles (640 million km) apart.

To help us understand just how far away Jupiter really is from Earth, a spaceship traveling at 55 miles an hour (88 km an hour) would not reach the planet for about 1,000 years! At the speed of light — 186,000 miles a second (300,000 km a second) — it would take about an hour. In reality, a spaceship cannot travel that fast, but radio waves do travel as fast as light. Scientists studying the planet have found that a radio signal beamed at Jupiter arrives there in 58 minutes. So, we can say that Jupiter is 58 "light-minutes" away from Earth.

► This cross section of Jupiter's outer atmosphere represents a region that is about 125 miles (200 km) thick. The temperature increases quickly from the top to the bottom. The upper clouds are at −270°F (−170°C), the **hydrogen** sulfide and ammonia crystal region is at −170°F (−110°C), and the dividing line between the ice crystals and water droplets is at around 32°F (0°C). Below the water-droplet clouds, the atmosphere of Jupiter merges with the underlying oceans of liquid hydrogen.

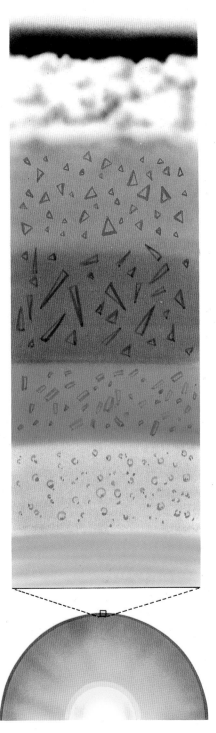

A vast atmosphere

Perhaps the best way to describe Jupiter is to compare it to a planetary body that we know well — Earth. Our home planet is a rocky world, partially covered with water, and surrounded by an atmospheric blanket. Jupiter is mostly atmospheric gas, with very little water, and possibly only a tiny rocky core under all those clouds.

Jupiter has a **volume** 1,330 times that of Earth. In fact, 11 Earths could be fitted across Jupiter's face. If we could drive in a car around Earth's equator at 55 miles an hour (88 km an hour), we would make the 25,000-mile (40,000-km) "round" trip in about 19 days, during which time we would see a lot of interesting geography. On Jupiter, the same journey would be 280,000 miles (450,000 km) and traveling at 55 miles an hour (88 km an hour) would take us 7 months. We would see no solid surface geography on Jupiter — just the cloud tops.

The jovian system

Earth has one moon, a rocky satellite 2,160 miles (3,476 km) in diameter, covered with craters and dark areas which we call "maria," or seas. Jupiter has at least 16 moons. Some are icy bodies, others are large chunks of rock.

One Earth day lasts 24 hours, because the planet spins on its axis and takes 24 hours to make one complete turn. (This is called the period of rotation.) Jupiter spins about twice as fast as Earth, and one day lasts just under 10 hours. It takes 4,330 Earth days to orbit the Sun because it is so much farther away and travels more slowly. The planet's gravity is around two and a half times that of the Earth's, and its **mass** is 317 times greater.

Canyons between the clouds

▼ Even deeper within one of the broad, white canyons of Jupiter's mid-latitude zones, the huge extent of the jovian weather systems can only be suggested by the detail shown in the foreground of the picture. Here, each little puff of cloud is many miles (km) across, equivalent in extent to one of Earth's large thunderclouds.

▲ Descending into one of the bright, white, planetary zones on Jupiter, the clouds lie in parallel banks with deep canyons between them. The immensity of the canyon is shown by the size of the distant cloud formations and the finer details in the foreground clouds. A thin haze of orange gas can be seen pouring into the lower region where winds are howling and clouds are scudding past.
The distant Sun is surrounded by an upper atmospheric haze.

The Great Red Spot

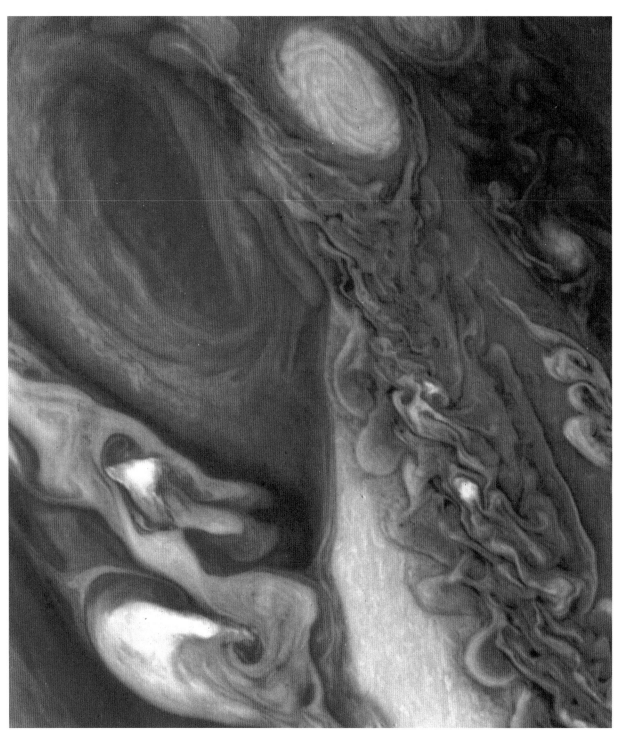

Jupiter, the gas giant

In the solar system, there are two types of planets — "terrestrial" worlds and "gas giants." The terrestrial worlds — Mercury, Venus, Earth and Mars — are rocky bodies. Gas giants are planets made up of gases and liquids with small rocky cores at their centers, namely Jupiter, Saturn, Uranus and Neptune. We use the word "small" because the cores only make up a small percentage of their total mass. The cores of Jupiter and Saturn may be 20 times more massive than the whole of planet Earth. A study of Jupiter is useful in giving us clues to the composition of the other gas giants as well.

Jupiter is the most impressive gas giant in the solar system. It dwarfs all the other planets in every way and accounts for around 70 percent of the total mass of the solar system.

A considerable part of Jupiter's mass consists of its atmosphere — an 800-mile (1,280-km) thick layer blanketing the planet. Hydrogen, helium, water and ammonia ice crystals form its gracefully sculpted clouds. Poisonous gases also exist within the clouds: methane, ammonia, ethane and acetylene. Large amounts of hydrogen are needed to form these gases, and Jupiter has hydrogen in several different forms throughout its atmosphere.

Lightning and storms

Jupiter rotates very fast. The rapid turning causes very high winds blowing at speeds up to 300 miles an hour (480 km an hour). The gases in the atmospheric clouds circulate around the planet, and the clouds are churned into the colorful belts and zones we see from Earth. Lightning results from this violent mixing of the atmosphere.

Diving into the depths

The cloud belts and zones around Jupiter all move at different speeds. Storms are created along the edges of the belts, where two different wind-speed regions meet. Winds in these boundary layers flow in a circular motion, very much like hurricanes on Earth. In many photographs of Jupiter's atmosphere, we see white and yellow storms dotted along the boundary zones.

One of the most famous of these storms is the Great Red Spot, which has been raging in Jupiter's atmosphere for at least 300 years. It is so large that two Earths could fit inside it. Jupiter's clouds are colorful and patterned, but what lies beneath them is a mystery. In future exploration of this gas giant, scientists will probably try to send probes into the depths of the atmosphere.

◄ This picture was taken by Voyager 1 as it flew past Jupiter. The Great Red Spot is a vast, swirling system of clouds, appearing like a planetary hurricane — the biggest storm in the solar system. The wind speeds can be as high as 310 miles an hour (500 km an hour). Discovered by the British scientist Robert Hooke in 1664, the storm has persisted for more than three centuries. The spot is much colder than its surroundings, suggesting that it is an elevated region of high pressure rotating counterclockwise and drawing material up from a region below the cloud level.

Material then falls back, around the edges of the spot, which is slowly shrinking and is now only half the size that it was a century ago.

Cloud mountains

▶ Cloud structures of towering majesty fill the skies, where the edge of a band of brighter white clouds thrusting above the lower cloud layers can be seen.

This region is about 50 miles (80 km) thick. The lowest clouds are made up of water ice crystals, or liquid raindrops, while the higher clouds are ammonium ice. Blue clouds are the deepest and hottest, becoming brown, then white and, finally, red as they get higher and cooler. The different colors are produced if some of the particles in the cloud are electrically charged, if lightning strikes the cloud, if the cloud is moving up quickly, or if the cloud has become saturated with **sulfur**.

Jupiter's structure

If we could cut a slice out of Jupiter we would find a layered structure, very different from our planet Earth.

A hot core lies at the center of the Earth, where rock is molten and flows like water. Above the core is the mantle, an area where rock is cooler, but still **viscous**, lying below Earth's rigid crust. When great pressure is applied to the crust, it breaks. Some parts of the crust are covered by oceans, and surrounding the entire planet is an atmosphere 100 miles (160 km) thick.

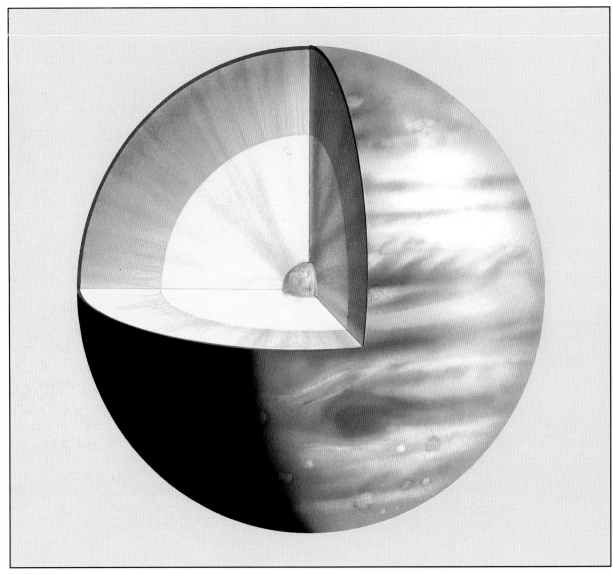

Jupiter, too, has a hot, rocky core — about 20 times the mass of the Earth. No one has mapped this core yet, but scientists think it is made of **silicon** and traces of iron. Hydrogen, water and gases surround the center like the layers of an onion. Jupiter's atmosphere is eight times thicker than Earth's.

A journey to the center

If we traveled down through Jupiter's atmosphere toward the center, we would encounter first, the upper cloud layers, which are divided into belts and zones. The zones range in color from light tan to a yellowish green. The belts are darker, and show browns, reds and even some green in their coloration. All the atmospheric gases mix, creating hazes very much like the smog we see on Earth. The temperature in these areas is cold — about −220°F (−140°C). Through openings in the clouds, we see farther down, into warmer, hazy areas of hydrogen and helium mixes. Ammonia and water ice crystals hang suspended in the atmosphere. Here, temperatures are only −9°F (−23°C).

As we descend through thicker clouds and warmer layers, the temperature rises enough to allow water droplets to condense. At this level, the mix of gases resembles the atmosphere of the Earth before oxygen and nitrogen were dominant. Temperatures are warm enough to support life, and some scientists have wondered if there might be bizarre jovian life-forms floating through these warm, wet regions. Unfortunately, if there are, winds blowing in up- and down-drafts would push any creatures from their comfortable nesting grounds into regions too hot or too cold for them to survive.

Farther down, the temperature and atmospheric pressure continue to rise until the hydrogen gas becomes a liquid in a realm of hydrogen oceans. Below those oceans, the liquid hydrogen atoms are **ionized**, and their **electrons** move about quickly, conducting electricity like a metal. This liquid metallic hydrogen covers Jupiter's rocky core.

While travel to these depths is not yet possible, learning about the interior composition of Jupiter can help us to understand the interior compositions of the other gas giants. Nowhere else in our solar system do atmospheric scientists have such a wide variety of cloud formations and gaseous mixtures to study, and with each encounter, they will be able to understand better the makeup of planets so different from the rocky, terrestrial Earth. Jupiter, especially, catches the interest of those who study planetary atmospheres. Within the brilliantly-colored jovian cloud zones and belts lie clues to the origin of the solar system itself. In the future, probes descending into Jupiter's atmosphere may enable researchers to uncover those clues, and unlock the mysteries of Jupiter's inner secrets.

◀ The thickness of the cloud zone around Jupiter is only about ⅟₅₀₀ of the radius of the planet. If we collected 1 million molecules from these clouds, 1 molecule would be water, 200 ammonium, 700 methane, 100,000 helium and the rest hydrogen. Beneath the clouds the atmosphere is clear extending downwards until it becomes an ocean of hydrogen and helium. At a depth of 10,500 miles (17,000 km), the liquid acts like a metal making up 77 percent of the total mass of the planet. At the top of the inner layer the temperature is more than 18,000°F (10,000°C), increasing to 36,000°F (20,000°C) at the bottom.

The rocky, metal core has a composition similar to Earth's, and is subjected to very high pressure.

Observing Jupiter

Observation from Earth

How do we know so much about Jupiter? Until the start of the Space Age, all our planetary observations were made from Earth. For centuries, people studied the planets with the naked eye, but with the invention of the telescope in the 1600s, views of the planets became clearer, enabling astute astronomers, like Galileo Galilei in the 17th century, to observe that Jupiter had satellites. As telescopes improved, so did the level of knowledge about the planet. Photography allowed Jupiter-gazers to capture images of the cloud bands, the Red Spot and the largest satellites.

However, Earth-based observations are affected by our atmosphere. If we look at Jupiter through a telescope, the image wavers and shimmers before our eyes, due to the movement of our atmosphere.

Observation from space

Since 1973, the National Aeronautic and Space Administration (NASA), with researchers and scientists from hundreds of universities and private companies, has launched a series of four spacecraft toward Jupiter, some traveling at speeds of up to 30,000 miles an hour (48,000 km an hour).

The spacecraft — Pioneer 10, Pioneer 11, Voyager 1 and Voyager 2 — are **robotic probes**, sent on one-way journeys past Jupiter and the outer planets. They are taking a "grand tour" of the gas giant worlds, sending back a steady stream of data for us to study. Operating on their own, with computer data radioed to them from Earth, they are self-powered, using electrical generating plants called **radioisotope thermal generators**. The craft are loaded with instruments which scientists use to study all aspects of the distant planets.

▼ Huge electrical storms often rage in the jovian atmosphere producing many bolts of lightning, which have been imaged by spacecraft passing over the night side of the planet. The storms only occur above certain regions whose positions seem to be influenced by the satellite Io, and produce intense radio wave emissions.

Orbiting satellites

Like our Sun, Jupiter keeps smaller bodies, its moons, in orbit around itself. The four largest are called the Galilean satellites, after Galileo, who discovered them. They are named Io, Ganymede, Callisto and Europa — mythological figures associated with Jupiter, the king of the gods.

The Galilean satellites have been most closely studied by the Voyager spacecraft teams, with startling results. Three of the four tiny worlds are icy bodies, with rocky interiors caused by both bombardment by **asteroids** and internal movements of their crusts and mantles. Their surfaces are scarred with craters, cracks and grooves. The fourth moon, Io, also has craters, but of a very different variety.

The remaining 12 satellites, discovered more recently, are Pasiphae, Elara, Amalthea, Himalia, Sinope, Adrastea, Ananke, Carme, Leda, Thebe, Lysithea and Metis. There are certain to be others even smaller, not yet discovered.

Jupiter's ring

Scattered between Jupiter's satellites are moonlets, each smaller than 125 miles (200 km) in diameter. These tiny chunks of rock are probably asteroids, captured by the planet's immense gravity, sweeping out paths in the dusty space. A ring of very fine particles also encircles Jupiter. Scientists think that this debris resulted from the collisions between rocks and particles, as well as from volcanic eruptions on Io. The ring is quite thin, probably no more than a few miles (km) across.

◀ In 1979, Voyager 1 discovered a faint ring around Jupiter, imaged in more detail by Voyager 2. The main band has a sharp edge 1.81 times the radius of the planet, fading gradually at 4,000 miles (6,400 km) nearer Jupiter. The ring is quite smooth and the brightness, although faint, is almost uniform. It lies in the same plane as the major satellites, and it is composed of particles only about 0.0002 inch (0.0005 cm) in size. Jupiter's ring continuously loses dust and is supplied with new particles, which have been eroded from the surface of large boulders orbiting in the same region.

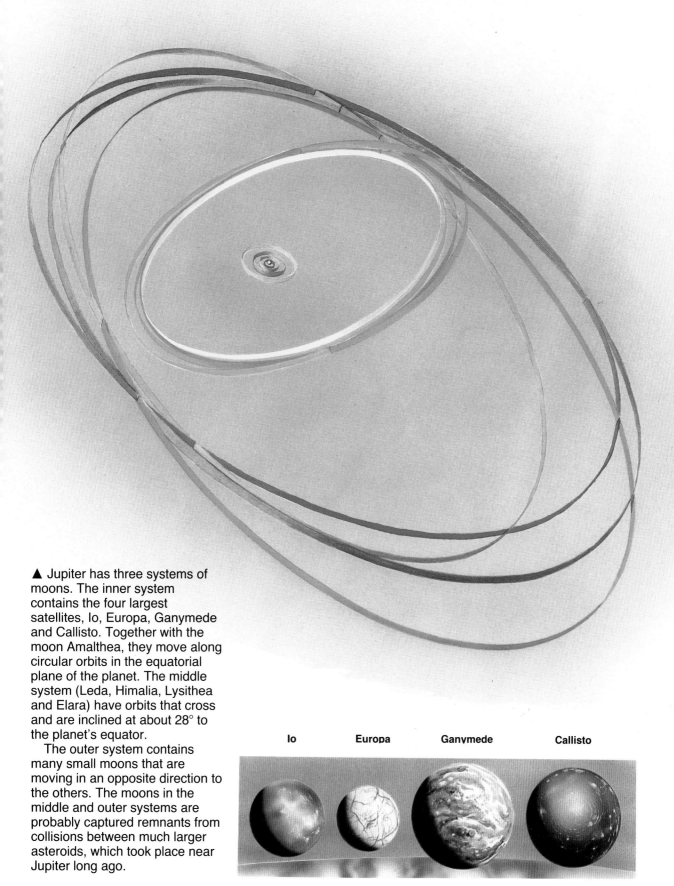

▲ Jupiter has three systems of moons. The inner system contains the four largest satellites, Io, Europa, Ganymede and Callisto. Together with the moon Amalthea, they move along circular orbits in the equatorial plane of the planet. The middle system (Leda, Himalia, Lysithea and Elara) have orbits that cross and are inclined at about 28° to the planet's equator.

The outer system contains many small moons that are moving in an opposite direction to the others. The moons in the middle and outer systems are probably captured remnants from collisions between much larger asteroids, which took place near Jupiter long ago.

| Io | Europa | Ganymede | Callisto |

21

Galilean moons in detail

The Galilean satellites are about the same size as the smaller planets in our solar system. Ganymede is a little bigger than Mercury and, apart from Europa, they are all bigger than Earth's Moon. Their orbits lie very close to the plane of Jupiter's equator and are all nearly circular. The satellites are less dense the farther they are from the central planet. A similar sequence of densities occurs in the solar system as one moves away from the Sun. Their surfaces show a fascinating variety of colors and textures.

Jupiter's magnetosphere

Jupiter has a large magnetic field, with a total strength about 4,000 times greater than the Earth's. This field dominates a region of space surrounding the planet with a radius of around 3.5 million miles (5.6 million km). Trapped in this region are many electrons and ionized atoms, which move in a spiral fashion along the field lines and bounce backward and forward between the two magnetic poles. The whole region is called a **magnetosphere** and everything within its reach is greatly affected by the existence of the magnetic field.

The Galilean satellites have orbits which lie deep within Jupiter's magnetosphere. They are continually bombarded by particles trapped by the field lines, causing discoloration and erosion. A considerable amount of material from Io is blasted from the surface and forms a donut-shaped ring of **plasma** around the planet. The plasma is hot and gives off **ultraviolet** light. It is possible that there is a massive electrical current connecting the poles of Jupiter with the satellite Io. Its power is thought to be about 70 times greater than the total generating capacity of all the nations on Earth combined, and could play an important role in heating certain regions on Io.

Facts about the Galilean moons				
	Io	Europa	Ganymede	Callisto
Mass (Moon =1)	1.213	0.663	2.027	1.448
Density (water =1)	3.53	3.03	1.93	1.79
Diameter miles	2,257	1,943	3,279	2,996
(km)	(3,632)	(3,127)	(5,277)	(4,822)
Orbital period (days)	1.77	3.55	7.16	16.69
Distance from Jupiter				
miles	262,000	417,000	665,000	1,170,000
(km)	(421,600)	(671,000)	(1,070,200)	(1,882,900)
Reflectivity	60 %	60 %	40 %	20 %

► Callisto is the most distant of the four Galilean satellites. Its surface is much older than Ganymede and is pockmarked with both ancient and new craters. It is like the more familiar landscapes of the inner solar system bodies, such as the Moon and Mercury, but highlighted with ice. Craters are scattered across its surface almost uniformly, but due to the thick, but weak, ice crust surrounding the moon, none of the crater walls are very high.

All the Galilean satellites orbit Jupiter in its equatorial plane. Ganymede, Europa and Io can be seen in the background, lined up with Jupiter's equator.

The satellites' origins

The densities of Io and Europa, measured by noting how the **gravitational field** of each satellite affects its neighbors, indicates that they are essentially rocky bodies. Ganymede and Callisto are less than twice the density of water, so they must have a large non-rock component which is most likely solid or liquid water.

Why does this difference exist? In the early days of the solar system, it is believed that Jupiter was so hot it was almost star-like. At that time its inner satellites were much warmer than its outer ones. The large cloud that surrounded Jupiter, from which the Galilean satellites condensed, probably contained equal masses of water and **silicate** rock. Energy lost when dust particles collided flattened out the satellites' orbits to form an equatorial disc. In the Ganymede–Callisto region it was cold enough for all the material to condense, but closer to Jupiter, it was so warm that only water-enriched rock could exist. The lighter elements either stayed in the cloud as gases, or were boiled off the surfaces of the condensing rocks rapidly.

23

Jupiter's nearest neighbors

◄ Our knowledge of Jupiter increased enormously when the two US spacecraft, Voyager 1 and Voyager 2, flew past in early March and early July 1979. The paths of the two craft were designed to provide the best opportunities for obtaining images of the major satellites. Both were able to take pictures of much of the surface of Io, because that satellite completed one orbit while the spacecraft were in the vicinity

This painting is the view from Voyager 1 of Jupiter and Io. The ring girdling Jupiter's equator is clearly visible, although the artist has shown it to be brighter than it really is. Many volcanoes dot Io's brightly colored surface — notice the one on the night side with just the top of its plume illuminated.

Satellite interiors

The interiors of the Galilean satellites can be divided into three regions. Io's crust, tens of miles (km) thick in places, is made of silicate rock containing considerable amounts of sulfur and **sulfur dioxide**. In fact, some scientists think that the outer part of this crust is a thin skin of solid sulfur floating on a vast ocean of molten sulfur. The interior of Io consists mainly of molten silicate rock, with a solid rock core about 600 miles (1,000 km) wide.

Europa is smaller and colder. The outer crust is solid ice and about 60 miles (100 km) thick. Beneath this crust is probably a region of slushy, dirty water ice. Again, the center consists of a molten silicate rock region surrounding a rocky core. Ganymede and Callisto both have icy crusts overlying hotter, liquid water oceans, 400 miles (640 km) deep, with large silicate cores.

▼ This photograph of Europa was taken from Voyager 1. Bright areas are probably deposits of ice, and dark patches may be rocky surfaces. The long, linear features cutting across the satellite's surface could be faults or fractures in the crust. Many of them are over 600 miles (1,000 km) long and 200 to 300 miles (300 to 500 km) wide.

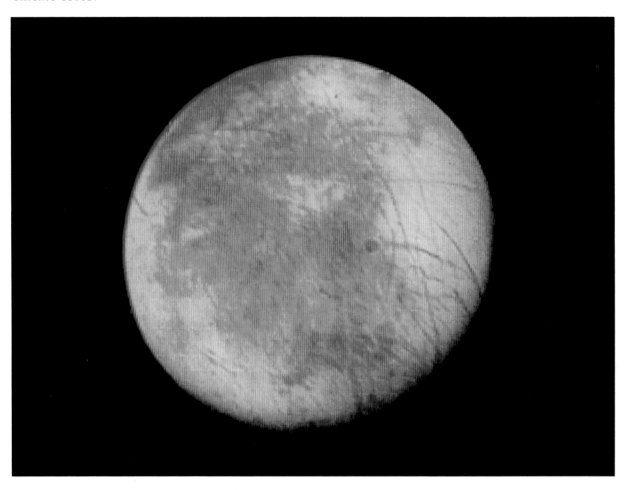

► In this Voyager 1 photograph, Io's unique orange color is displayed in striking detail. The smallest features shown here are about 6 miles (10 km) across. Patches of potassium, sodium and sulfur salts may explain the mottled surface, possibly derived from volcanic activity.

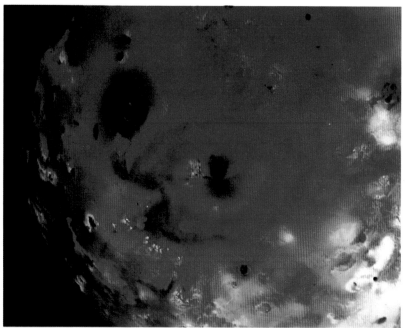

Heat sources

To maintain its molten interior, Io needs a constant source of heat. The main energy source on Earth, Mars and the Moon is the heat emitted when radioactive rocks decay. This source decreases with time until it eventually becomes negligible. Most small satellites radiate away all their heat and cool off, but Io is an exception.

Since Io is in close proximity to Jupiter, the planet's gravitational field exerts a large pull on Io forming a **tidal** bulge. Usually Io is in a **synchronous** orbit, where it takes the same time to spin around its axis as it does to complete one orbit, and the tidal bulge remains in the same spot. But the opposing effect of the gravitational pull of the other Galilean satellites often forces Io into an **elliptical** orbit and the bulge moves across its surface. The friction caused by this movement is given off as heat.

Another source of heat on Io is the jovian magnetosphere. Jupiter takes almost 10 hours to spin once around its axis, while Io orbits every 1.77 days. The magnetic field of Jupiter rushes past Io at a speed of 35 miles a second (57 km a second), generating an electrical force of 600 kilovolts across Io, with electrical currents of 1 million amps flowing from Jupiter to its satellite. The amount of energy generated between Jupiter and Io every second is the same as the total consumption of electricity by Western Europe in one year! Some of the power in the current flow that is dissipated in Io is a very effective source of heat. As a result of this massive input of energy, Io has many molten regions and is by far the most volcanically active body in the solar system.

Io

Io is the closest Galilean moon to Jupiter, and one of the most unusual bodies in the solar system. It is about the size of Earth's Moon, and apart from Earth, Io is the only place we know to have active volcanoes. When Voyager 1 flew by Io in March 1979, eight huge volcanic plumes could be seen. Four months later, as Voyager 2 passed by, six of these were still active and two new volcanoes had erupted.

A volcanic world

Why do volcanoes exist on a small moon in orbit around Jupiter? Io is "locked" to Jupiter by the planet's strong gravitational pull. The other Galilean moons also orbit the planet, though farther away, and the gravity of one — Europa — pulls Io away from Jupiter. Io is being tugged in two directions and the forces exerted heat up its core. The surface cracks and hot fluids from the underlying molten regions push their way up to the surface, forming volcanoes. Scientists call this effect tidal heating. Io is not the only body in the solar system to be affected by tidal heating, but its high sulfur content makes it the most unusual and dramatic.

▶ In the foreground is the cindery surface of Io, a satellite 1.2 times bigger than our Moon, with about the same density. Io is mostly red and orange and it has violent active volcanoes on its surface, ejecting lava at speeds of around 0.6 miles a second (1 km a second). The lava plumes rise up to 60 miles (100 km) high before fanning out to form an umbrella-shaped cloud and falling back to the surface. The erupting **magma** is a sulfur-rich silicate.

The surface of Io is covered with many overlapping lava flows. The volcanic debris appears blue when seen against the blackness of space, and brown against the bright, planetary background.

The swirling, jovian clouds decorate the huge, flattened globe with great varieties of color and texture. On the night side of the planet the flashes of scattered lightning storms and the glow of the polar aurorae can be seen.

◀ As Io is an active satellite, the finer surface details can change over short time periods. This view of the south pole shows an area that has few well-defined features. Io is thought to have a crust about 15 miles (20 km) thick, and the tidal forces exerted by Jupiter make it rise and fall by about 60 miles (100 km) twice every 1.8 days. The friction generated by this process is released as heat, which melts the volcanic magma.

The colorful moon

Io's surface

Io is extremely colorful, with its orange, red and black surface. If we could stand on this moon, we would see Jupiter through a yellow sulfur "snowstorm." Lakes of sulfur lava dot the surface, and plumes of sulfur particles jet into the atmosphere. Io's face shows many different colors — ranging from yellow, to red, black and green. The color of the element sulfur, which makes up a major part of Io, changes with the temperature.

Io is a long way from the Sun, so the normal daytime temperature is around −230°F (−145°C). Scattered on the surface are warm spots where the crust is weak. Here, the temperature hovers around 80°F (30°C). The active volcanoes can reach temperatures of over 620°F (330°C), much higher than the melting point of sulfur at 234°F (112°C), but considerably cooler than the molten silicate lava that erupts from volcanoes on Earth.

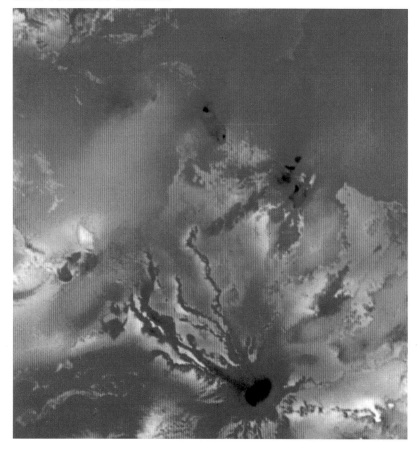

◀ This close-up view of Io's surface, captured by Voyager 1's cameras, shows an area about 600 miles (1,000 km) across. The most striking feature is the dark patch at the center surrounded by a radiating pattern very much like the lava flows around an Earth volcano. Similar features also reveal craters like Earth's calderas.

► Another Voyager photograph of Io's surface has captured a volcanic eruption in progress on the moon's **limb**. Material is being ejected to about 100 miles (160 km) above the surface at almost 1,200 miles an hour (2,000 km an hour).

Volcanic caldera

On other moons in the solar system, craters have been formed by collisions with smaller objects. The craters found on Io though, are volcanic **caldera**. The lava from these volcanoes has covered the moon's surface. The impact craters produced by collisions are submerged beneath lava flows in a few hundred thousand years. The volcanoes on Io are so active that it has probably completely turned its surface layers "inside out" at least once since the solar system formed. Thus, Io has a relatively young surface, since it is constantly covered and recovered.

Europa

After the volcanic activity on Io, Europa seems a tame, quiet world. It is the smoothest object in the solar system, almost as featureless as a billiard ball. Except for a slight rise here and there and the remains of an ancient crater, the terrain is nearly flat. Yet its smooth, cream-colored surface of water ice is crisscrossed with tan-colored cracks, 40 miles (65 km) across, and over 600 miles (1,000 km) long. The key to their origin can be found beneath the surface.

Scientists think that an ocean of slushy water lies under Europa's glacier-like 62-mile (100-km) thick ice crust, and beneath that ocean almost certainly is a cool, rocky core. To have a slushy ocean, however, a moon needs a source of heat, just to keep the water from freezing solid. Unfortunately, there is not enough heat coming from the core to do this, and so we must look at Europa's location in the jovian system. It is outside the orbit of Io, but it is still close enough to be affected by jovian tides.

The moon Ganymede also pulls Europa away from Jupiter, and like Io, Europa's interior is warmed by tidal heating. Since Europa is composed mostly of water ice, that ice melts and the dirty water rushes to the surface, like lava in a volcano. The stress on the moon from the stretching and pulling causes cracks on the surface, and water rushes up to fill them in. When it reaches the surface, it freezes into the lines and grooves seen here.

A few impact craters can also be seen on Europa's surface but over the years these craters have been covered with ice, or their walls have slumped down under its weight. Europa's surface, like Io's, is relatively young.

▶ In this painting, Jupiter is seen from Europa, the icy satellite in the foreground. Europa is three times more dense than water and is the smallest of the Galilean moons, being only ⁹/₁₀ the size of our Moon. The surface is completely different from that of its neighboring satellite, Io, which is closer to Jupiter. Io is a reddish color, Europa white; Io is volcanically active, while Europa is dead.

Europa resembles a smooth, white, reflecting sphere. Nearly crater-free, it is crisscrossed by a confusing maze of bright and dark lines, which are nearly straight, about 25 miles (40 km) wide and thousands of miles (km) long. They are thought to be cracks in the icy surface filled with dirty ice. There are also ridge-like creases crossing the surface, around 300 feet (90 m) high, 5 miles (8 km) wide and 60 miles (100 km) long.

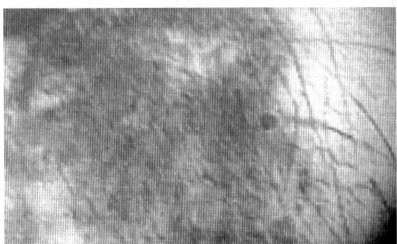

◀ The thick icy surface of Europa may cover an ocean of water. Occasionally, the cracks may split open and water seeps onto the surface and freezes.

Ganymede

Farther out from Io and Europa is the moon Ganymede, the largest satellite in the solar system. Like Europa, it consists mostly of ice, with a tiny rocky core.

Ganymede's surface is wrinkled, covered with dark brown soil and lighter white-gray ice. The dark areas have many craters, the wrinkles being the remnants of very old, multiringed features. The bright bands show grooved areas, but the fact that there are few craters indicates that the satellite has been resurfaced.

Ganymede may have formed with a very thin icy crust — perhaps only 6 miles (10 km) thick. As with Europa, there may be a slushy ocean under the ice, but the effect of tidal heating is less than on Io and Europa.

Scientists studying this small moon think that large blocks of ice may drift in the under-surface ocean. Occasionally these blocks may crack the surface and water gushes up through the fissures, almost like lava from a volcano. The water then freezes, filling the cracks with fresh, new ice.

Ganymede has two extensive polar caps extending to latitudes of about 40° and its surface has broken up into plates, like the continents on Earth. They have only moved a few tens of miles (km), however, so this jostling has not led to the creation of chains of mountains like those on Earth.

▼ Like the majority of moons, Ganymede is small, cold, and has no atmosphere, but its surface is most unusual, scored with many long straight grooves and dotted with impact craters surrounded by bright ray patterns.

▶ The crescent Jupiter can be seen hovering in the sky above the icy surface of Ganymede.

In this view, we are looking from the darker, more heavily cratered highlands along one of these grooves, which is about 6 miles (10 km) wide and quite shallow. Ganymede, like its neighbor Europa, is a smooth satellite, much smoother than our Moon. Its icy surface has become dirty with age, but when a new crater is formed by an impacting asteroid, the fresh, nearly white underlying ice is splashed over the surface. The craters formed on the icy surface of Ganymede have a different profile to those in the rocky surface of the Moon, being less deep.

▼ This close-up photograph of Ganymede's surface was taken by Voyager 1 at a range of 158,000 miles (254,000 km) and shows an area 600 miles (1,000 km) across. Ths smallest features are 1.5 miles (2.5 km) wide.

Callisto

Callisto is a world pockmarked with craters. Because these craters are still visible, we can conclude that its surface is ancient, not constantly covered over with new material, as on the young surfaces of Io, Europa and Ganymede. Like the other moons, Callisto is icy, with a hard, rocky core. It orbits at a distance of 1,170,000 miles (1,882,000 km) from the cloud tops of Jupiter, as it is the farthest of the Galilean satellites.

Beneath its icy crust, Callisto is probably like the other moons — a water, or slush, ocean surrounding a tiny rocky core. There is no moon farther out to change its circular orbit so the tidal heating is minimal. Its surface has not been recovered with lava or slush, but has been bombarded with rock since its formation.

A cratered crust

Callisto's surface is dark and covered with soil. Small **meteorites** have hit the surface and the energy of these impacts has vaporized the ice, leaving the dusty soil behind. Water, which sprayed out as a result of collisions, has drenched the surrounding areas with bright rays of ice. The craters are flat, because they have been formed in dirty ice. The ice has then flowed like a glacier and leveled out the crater walls, leaving behind scars shaped like rings. Ripple rings surround Valhalla, the largest crater on the moon, and stretch across nearly one-quarter of the satellite. Smaller craters have smaller rims, so the ice is rigid enough to support them.

► Callisto's surface, photographed by Voyager 1, displays many features similar to those of the Moon and Mercury. There is heavy impact cratering with the more recent craters surrounded by bright ray patterns. The enormous number of craters suggests that Callisto is the oldest of the Galilean moons.

◄ This photograph of the giant Valhalla crater was taken by Voyager 1. It shows clearly the circular ripples, or rays, surrounding the impact site.

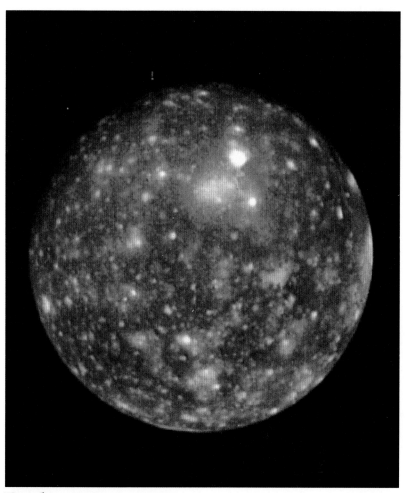

Twelve more moons

Jupiter's remaining satellites, in the middle and outer groups,
are so small and faint that they could not be seen through
small telescopes and were only discovered from photographic
images. They all seem to be either captured asteroids or
remnants of a jovian satellite that has been hit by an asteroid.

Jupiter's other moons					
Moon	**Diameter in miles (km)**	**Distance from Jupiter***	**Moon**	**Diameter in miles (km)**	**Distance from Jupiter***
Metis	25 (40)	0.080 (0.128)	**Lysithea**	16 (25)	7.325 (11.72)
Adrastea	16 (25)	0.081 (0.129)	**Elara**	50 (80)	7.335 (11.74)
Amalthea	131 (210)	0.113 (0.181)	**Ananke**	16 (25)	13.250 (21.20)
Thebe	63 (100)	0.138 (0.222)	**Carme**	19 (30)	14.125 (22.60)
Leda	6 (10)	6.934 (11.09)	**Pasiphae**	25 (40)	14.688 (23.50)
Himalia	113 (180)	7.175 (11.48)	**Sinope**	19 (30)	14.813 (23.70)
Note* in million miles (million km)					

Pioneer and Voyager

Pioneer 10 was the first unmanned probe to be launched into the solar system, leaving Earth orbit in 1972, and arriving for a **flyby** of Jupiter in December 1973. Pioneer 11 was launched in 1973 and swept past Jupiter in December 1974. As it passed the planet, Pioneer 11 received a **gravity boost** from Jupiter, which increased its speed and sent it on its way to a flyby of Saturn in 1979.

The Pioneer missions

Both Pioneer spacecraft shared the same goals: to navigate the asteroid belt, and to measure the level of radiation around Jupiter. The asteroid belt crossing was successful, and the radiation tests were all completed, although the spacecraft electronics were affected as they passed through the very strong radiation belts contained in Jupiter's magnetosphere. Scientists used the data about the radiation hazards encountered to redesign the electronics on the Voyager spacecraft.

The Pioneer spacecraft are still operating and transmitting data — more than 15 years after their departure from Earth! Their paths are carrying them out of the solar system, and they will wander the depths of interplanetary space forever.

The Voyager missions

Voyagers 1 and 2 were launched in 1977, Voyager 2 on August 20 and Voyager 1 on September 5, and both flew by Jupiter in 1979, before heading for Saturn.

Each probe carries 11 instruments including video cameras, antennae and instruments to study the planet's magnetic fields, examine its atmosphere and listen to Jupiter's loud and powerful radio "voice." On-board computers control most of Voyager's motions and activities, but some commands from Earth are still needed. All the instruments and computers are tightly packed, and the whole spacecraft weighs less than a small car. The Voyagers use the same type of radioactive generating plant to supply electricity as the Pioneers, but use less power than a 100-watt light bulb!

▶ Pioneers 10 and 11 gave us our first close-up views of Jupiter in 1973 and 1974. Pioneer 10 is pictured here on December 3, 1972, only 81,000 miles (130,300 km) from Jupiter, shooting past the planet on its way out of the solar system. In 110,000 years it will be 4 light-years away, in the vicinity of the nearby stars.

The large, **high-gain** dish antenna is pointing towards Earth, the **booms** extending from the spacecraft carrying the radioactive **thermoelectric** power generators. One of the most interesting experiments on board produced close-up pictures of Jupiter. The initial images were taken through two color filters, one blue and the other red. The final color pictures were produced by **computer enhancement** at the University of Arizona.

Galileo

Receiving the data

In order to communicate with the probes, a team of over 100 scientists, assembled at the Jet Propulsion Laboratory in Pasadena, California, use computers and satellite links. Because a one-way message takes more than an hour to reach the probes at Jupiter, scientists have to be very careful about the commands they send, as it takes too long to correct mistakes.

When a probe begins a "close approach" (that is, when it flies by the planet), the encounter with the planet usually lasts only a few days. This is one of the busiest and most exciting times for planetary scientists, when data streams in, in very large quantities. "Imaging technicians" work to produce the photographs, or images, of the planet which are examined by scientists.

The Galileo probe

It may be years before we understand everything that the Voyagers have revealed to us about the giant planet. Meanwhile, another spacecraft is being prepared to fly to Jupiter — the Galileo probe. Like the Voyagers, it will carry a full load of sensitive instruments and cameras, but Galileo will be launched from the US space shuttle, rather than by unmanned booster rockets. From an Earth orbit, it will move out beyond Mars, through the asteroid belt, and on to a **rendezvous** with Jupiter.

There, it will release a probe into the depths of the jovian clouds, which will report on atmospheric conditions before being crushed by the weight of the atmospheric pressure above it. Galileo then will enter into orbit around Jupiter, spending several years mapping the planet and studying the moons Ganymede, Callisto, Io and Europa.

The future

Perhaps in the future, spacecraft with multinational crews will orbit Jupiter to study its incredible wonders at firsthand. Until then, scientists will be occupied in examining and explaining the Voyager and Galileo data, in an attempt to unlock more of the mysteries of the gas giant, Jupiter.

▲ The Pioneer and Voyager missions were both flybys, with the spacecraft taking as many pictures as they could while they zoomed past.

The Galileo spacecraft will travel through space to Jupiter, where its **retro-rockets** will slow it down. It will fall into orbit around the planet, staying in Jupiter's vicinity forever. Galileo will fly by each of the satellites in turn, many times. The study of Jupiter itself is planned to last at least four years.

One of the most exciting aspects of the Galileo mission is the probe that will be launched into Jupiter's atmosphere. Here, the probe is floating under a parachute to slow its descent through the jovian clouds, measuring their temperature, density, pressure, wind speed and composition. Eventually, it will splash into the liquid oceans of the planet. Information from the descending probe will be sent back to the orbiting craft, which will then transmit it back to Earth.

Glossary

asteroids: a large group of minor planets that orbit the Sun between Mars and Jupiter. The largest is around 1,000 miles (1,600 km) across, the smallest only a few feet (m) in diameter.

aurorae: colored lights which shine in the sky caused when electrically-charged particles interact with gases in the atmosphere.

booms: long beams attached to the spacecraft.

caldera: the crater at the summit of a volcano.

computer enhancement: a photographic image from cameras on the space probe is intensified, or sharpened, by processing the data in a computer.

density: the mass of a substance divided by its volume (i.e. iron is twice as dense as rock and rock is 3.3 times as dense as water, so water has a lower density than rock).

disc: as planets are much closer to us than stars, you can see them as small discs when you look at them through a telescope. Stars look like points.

electrons: tiny particles of matter in orbit around the nucleus (center) of an atom. They have a negative electric charge.

elliptical: oval-shaped.

flyby: a space mission where the craft goes straight past the planet and does not slow down or fall into an orbit around it.

gravitational field: the space surrounding a body within which its gravity, the force that pulls objects towards each other, affects other bodies.

gravity boost: the gravitational field of the planet is used to speed up the spacecraft and give it the energy to go on to visit other planets.

high-gain: a radio aerial which amplifies signals received from, and transmitted to, Earth.

hydrogen sulfide: a compound of hydrogen and sulfur.

ionize: to convert into ions (groups of atoms that have a positive or negative electric charge).

limb: the edge of the visible disc.

magma: liquid rock that flows out from under the surface of a planet or moon.

magnetosphere: the region affected by the magnetic field of a body.

mass: the quantity of material in an object.

meteorite: the remains of a large meteor that has not been completely burned up as it falls through the Earth's atmosphere. It falls to Earth as a mass of metal or stone.

oblate: flattened at the poles.

plasma: a highly ionized gas containing equal numbers of positively charged and negatively charged ions. The Sun and the stars are made of plasma.

radioisotope thermal generator: an electrical power source which uses heat generated in a quantity of radioactive atoms.

rendezvous: to meet at an appointed place and time.

retro-rockets: a small rocket motor on a spacecraft for slowing it down or changing its orbit.

robotic probe: a probe designed to carry out measurements of the local environment and transmit its findings to Earth.

silicate: a salt derived from silica, a hard crystalline mineral substance found in flint, quartz, and sand.

silicon: non-metallic element found in silica.

sulfur: a hard, brittle, yellow substance that burns with a blue flame and forms an unpleasant-smelling gas.

sulfur dioxide: a heavy gas with a sharp, choking smell and no color. We use it as bleach and to preserve food.

synchronous: the same time is taken by a planet or moon to spin once around its axis as to complete one orbit.

thermoelectric: electricity produced by differences in temperature.

tidal: the effect produced by the distortion of a planet's, or satellite's surface caused by the gravitational pull of a nearby object.

ultraviolet: radiation with a shorter wavelength than the violet light we see. Ultraviolet rays occur naturally in sunlight and are invisible.

viscous: describes a liquid that does not flow easily. Oil and treacle are viscous liquids.

volume: the space taken up by a solid object, liquid or gas in three dimensions.

Mercury

Venus

Earth

Mars

◀ The Sun and the nine planets compared, with their sizes depicted on the same scale. In comparison to the planets, the Sun is so large that only a small part of it can be shown.

Jupiter

SUN

Saturn

Uranus

Neptune

Pluto

Index

Acknowledgments

ILLUSTRATIONS
COVER, 7, 8, 10, 11, 14, 15,
18, 19, 23, 24, 25, 28, 29,
33, 35, 39, 40, 41: Don Davis.
6, 9, 21: Sebastian Quigley/
Linden Artists.

PHOTOGRAPHIC CREDITS
12, 20, 26, 27, 30, 31, 32,
34 right and left, 36: NASA.
37: NASA/Science Photo Library.
43: BLA.